Preface

We are delighted to present the "Science Olympiad All Chapters Combined MCQ Class 3", a resourceful and engaging book designed to prepare students for Science Olympiad examinations. This book features Multiple-Choice Questions (MCQs) that comprehensively cover all chapters of the Class 3 science syllabus.

Science is a fascinating subject that encourages curiosity, discovery, and problem-solving. This book is crafted to help students build a strong foundation in scientific concepts while making learning enjoyable and interactive. Each question is tailored to align with the Olympiad exam format, ensuring students gain familiarity with the style and difficulty of the competition.

We hope this book serves as a valuable companion for Class 3 students, inspiring them to excel in Science Olympiad competitions and instilling a lifelong passion for science

Table of Contents

Chapter 1: Questions : Part 1

1. What do plants need to make their food?

a) Soil

b) Water

c) Sunlight

d) All of the above

2. Which part of the plant absorbs water and minerals from the soil?

a) Leaf

b) Stem

c) Root

d) Flower

3. What is the process called when plants make their food?

a) Digestion

b) Photosynthesis

c) Respiration

d) Transpiration

4. Which gas do plants take in from the air to make food?

a) Oxygen

b) Nitrogen

c) Carbon dioxide

d) Hydrogen

5. What is the main source of energy for the Earth?

a) Wind

b) Water

c) The Sun

d) The Moon

6. What do animals need to survive?

a) Water

b) Food

c) Shelter

d) All of the above

7. What do humans use to breathe?

a) Lungs

b) Heart

c) Stomach

d) Liver

8. Which of these is a living thing?

a) Rock

b) Tree

c) Water

d) Air

9. What do we call a scientist who studies animals?

a) Botanist

b) Zoologist

c) Chemist

d) Physicist

10. Which planet is known as the Red Planet?

a) Earth

b) Venus

c) Mars

d) Jupiter

11. What is the hardest natural substance on Earth?

a) Gold

b) Diamond

c) Iron

d) Silver

12. What do bees collect from flowers?

a) Water

b) Pollen

c) Soil

d) Seeds

13. Which part of the plant is usually underground?

a) Leaf

b) Stem

c) Root

d) Flower

14. What kind of animal is a frog?

a) Mammal

b) Reptile

c) Amphibian

d) Bird

15. Which organ is responsible for pumping blood in the human body?

a) Brain

b) Heart

c) Liver

d) Kidney

16. What type of animal lays eggs?

a) Mammal

b) Bird

c) Fish

d) All of the above

17. How many legs do insects have?

a) Four

b) Six

c) Eight

d) Ten

18. Which planet is closest to the Sun?

a) Venus

b) Earth

c) Mercury

d) Mars

19. What is the boiling point of water?

a) 50°C

b) 75°C

c) 100°C

d) 125°C

20. What do we call the young one of a butterfly?

a) Tadpole

b) Caterpillar

c) Larva

d) Nymph

21. What is the main gas found in the air we breathe?

a) Oxygen

b) Nitrogen

c) Carbon dioxide

d) Hydrogen

22. What is the process by which water vapor changes into liquid water?

a) Evaporation

b) Condensation

c) Precipitation

d) Sublimation

23. Which sense organ helps us to see?

a) Nose

b) Eyes

c) Ears

d) Skin

24. What do we call the movement of the Earth around the Sun?

a) Rotation

b) Revolution

c) Orbit

d) Tilt

25. Which planet is known for its rings?

a) Mars

b) Jupiter

c) Saturn

d) Neptune

26. Which of these is not a part of the water cycle?

a) Evaporation

b) Condensation

c) Precipitation

d) Combustion

27. What do we call animals that eat only plants?

a) Carnivores

b) Herbivores

c) Omnivores

d) Insectivores

28. Which material is attracted to magnets?

a) Plastic

b) Wood

c) Iron

d) Glass

29. What type of energy does a flashlight use?

a) Solar energy

b) Electrical energy

c) Wind energy

d) Kinetic energy

30. What kind of animal is a shark?

a) Mammal

b) Bird

c) Fish

d) Reptile

31. What do we call a baby cat?

a) Puppy

b) Kitten

c) Cub

d) Calf

32. Which part of the human body is made up of bones?

a) Skin

b) Hair

c) Skeleton

d) Muscles

33. What do plants release into the air during photosynthesis?

a) Carbon dioxide

b) Oxygen

c) Nitrogen

d) Hydrogen

34. Which planet is known as the Earth's twin?

a) Mars

b) Venus

c) Mercury

d) Jupiter

35. What is the largest organ in the human body?

a) Heart

b) Liver

c) Skin

d) Brain

36. What do we call the bones in our head?

a) Rib cage

b) Skull

c) Spine

d) Femur

37. What kind of animal is a lizard?

a) Mammal

b) Bird

c) Fish

d) Reptile

38. What is the center of an atom called?

a) Electron

b) Neutron

c) Proton

d) Nucleus

39. Which part of the plant is responsible for reproduction?

a) Leaf

b) Stem

c) Root

d) Flower

40. What do we call the path of an object as it moves around another object in space?

a) Rotation

b) Revolution

c) Orbit

d) Axis

41. Which planet is the largest in our solar system?

a) Earth

b) Venus

c) Mars

d) Jupiter

42. What is the main function of leaves?

a) Absorb water

b) Support the plant

c) Make food

d) Reproduce

43. What do we call a scientist who studies plants?

a) Zoologist

b) Botanist

c) Chemist

d) Physicist

44. What is the freezing point of water?

a) 0°C

b) 10°C

c) 20°C

d) 30°C

45. What is the smallest unit of life?

a) Atom

b) Molecule

c) Cell

d) Organ

46. Which of these is not a renewable energy source?

a) Solar energy

b) Wind energy

c) Fossil fuels

d) Hydro-power

47. What do we call the layer of air that surrounds the Earth?

a) Lithosphere

b) Hydrosphere

c) Atmosphere

d) Biosphere

48. What type of animal is a penguin?

a) Mammal

b) Bird

c) Fish

d) Reptile

49. What do we call the liquid rock that comes out of a volcano?

a) Lava

b) Magma

c) Ash

d) Crust

50. Which sense helps us to smell?

a) Sight

b) Hearing

c) Taste

d) Smell

Chapter 2: Questions : Part 2

51. What do we call the star at the center of our solar system?

a) Earth

b) Moon

c) Sun

d) Mars

52. What type of energy is stored in food?

a) Kinetic energy

b) Potential energy

c) Chemical energy

d) Thermal energy

53. What do we call a scientist who studies rocks?

a) Botanist

b) Geologist

c) Chemist

d) Zoologist

54. What is the largest planet in our solar system?

a) Earth

b) Venus

c) Mars

d) Jupiter

55. Which planet is known as the Blue Planet?

a) Mars

b) Earth

c) Neptune

d) Uranus

56. What do we call the bones in our back?

a) Skull

b) Spine

c) Rib cage

d) Pelvis

57. Which animal is known as the King of the Jungle?

a) Tiger

b) Lion

c) Elephant

d) Bear

58. What is the process by which plants lose water through their leaves?

a) Respiration

b) Transpiration

c) Photosynthesis

d) Absorption

59. What do we call the force that pulls objects toward the Earth?

a) Magnetism

b) Friction

c) Gravity

d) Electricity

60. What is the main purpose of the stem in a plant?

a) Make food

b) Absorb water

c) Support the plant

d) Reproduce

61. Which planet is known as the Earth's twin?

a) Mars

b) Venus

c) Mercury

d) Jupiter

62. Which part of the human body is used for thinking?

a) Heart

b) Brain

c) Stomach

d) Liver

63. What is the name of our galaxy?

a) Andromeda

b) Milky Way

c) Sombrero

d) Whirlpool

64. What is the main gas found in the air we exhale?

a) Oxygen

b) Nitrogen

c) Carbon dioxide

d) Hydrogen

65. What do we call the young one of a frog?

a) Tadpole

b) Caterpillar

c) Larva

d) Nymph

66. Which animal is known for changing its color to blend with its surroundings?

a) Frog

b) Chameleon

c) Snake

d) Lizard

67. What is the main function of roots?

a) Make food

b) Support the plant

c) Absorb water and minerals

d) Reproduce

68. What is the most abundant gas in the Earth's atmosphere?

a) Oxygen

b) Nitrogen

c) Carbon dioxide

d) Hydrogen

69. What do we call animals that eat both plants and animals?

a) Herbivores

b) Carnivores

c) Omnivores

d) Insectivores

70. What type of rock is formed from molten lava?

a) Sedimentary

b) Metamorphic

c) Igneous

d) Fossil

71. What is the smallest bone in the human body?

a) Femur

b) Rib

c) Stapes

d) Skull

72. Which planet has the most moons?

a) Earth

b) Mars

c) Jupiter

d) Saturn

73. What type of animal is a whale?

a) Fish

b) Reptile

c) Mammal

d) Amphibian

74. Which sense helps us to hear?

a) Sight

b) Hearing

c) Taste

d) Smell

75. What do we call the liquid part of our blood?

a) Plasma

b) Platelets

c) Red blood cells

d) White blood cells

76. What is the main function of the heart?

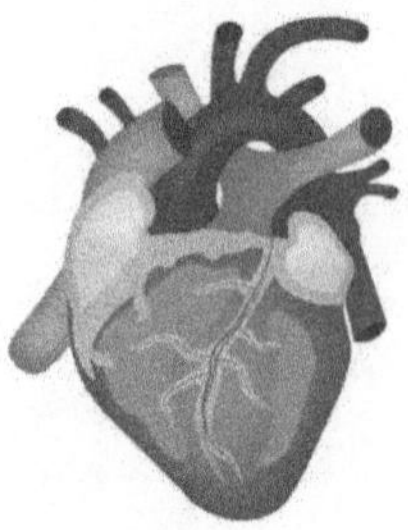

a) Digest food

b) Pump blood

c) Control movement

d) Filter waste

77. Which of these is not an insect?

a) Butterfly

b) Spider

c) Ant

d) Bee

78. What is the largest land animal?

a) Lion

b) Elephant

c) Giraffe

d) Rhino

79. What do we call the movement of water from the ground to the sky and back again?

a) Water cycle

b) Carbon cycle

c) Nitrogen cycle

d) Oxygen cycle

80. What do we call a baby dog?

a) Kitten

b) Cub

c) Puppy

d) Calf

81. Which of these is a nonliving thing?

a) Tree

b) Dog

c) Rock

d) Bird

82. What is the main purpose of flowers in a plant?

a) Absorb water

b) Support the plant

c) Reproduce

d) Make food

83. What type of energy comes from the Sun?

a) Electrical energy

b) Kinetic energy

c) Solar energy

d) Potential energy

84. Which planet is known for its Great Red Spot?

a) Earth

b) Jupiter

c) Mars

d) Saturn

85. What do we call a scientist who studies the weather?

a) Biologist

b) Geologist

c) Meteorologist

d) Zoologist

86. What is the main function of the liver?

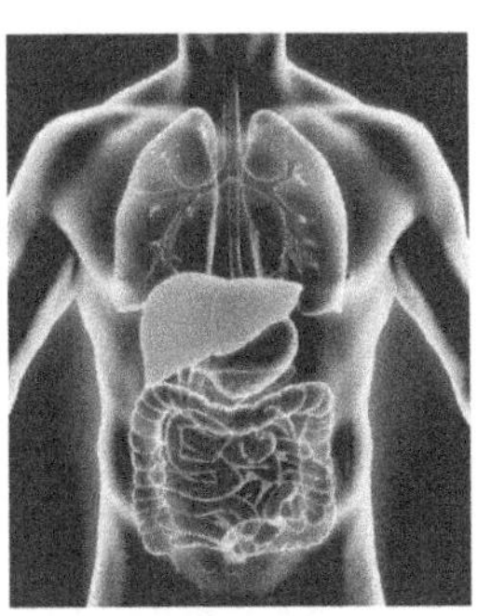

a) Pump blood

b) Filter waste

c) Digest food

d) Control movement

87. Which of these animals is coldblooded?

a) Dog

b) Bird

c) Snake

d) Elephant

88. What do we call the study of the stars and planets?

a) Biology

b) Geology

c) Astronomy

d) Meteorology

89. Which part of the plant is responsible for making seeds?

a) Leaf

b) Stem

c) Root

d) Flower

90. What type of animal is a dolphin?

a) Fish

b) Reptile

c) Mammal

d) Bird

91. What is the main function of red blood cells?

a) Fight infection

b) Carry oxygen

c) Clot blood

d) Provide nutrients

92. What do we call the smallest part of an element?

a) Atom

b) Molecule

c) Cell

d) Organ

93. Which part of the human body helps us to digest food?

a) Heart

b) Lungs

c) Stomach

d) Brain

94. What is the natural satellite of the Earth?

a) Sun

b) Mars

c) Moon

d) Venus

95. Which of these is a form of precipitation?

a) Fog

b) Snow

c) Wind

d) Cloud

96. What do we call animals that hunt other animals for food?

a) Herbivores

b) Carnivores

c) Omnivores

d) Insectivores

97. Which part of the human body is used for tasting?

a) Nose

b) Tongue

c) Ears

d) Eyes

98. What type of rock is formed from layers of sediment?

a) Igneous

b) Metamorphic

c) Sedimentary

d) Fossil

99. What is the main function of white blood cells?

a) Carry oxygen

b) Fight infection

c) Clot blood

d) Provide nutrients

100. Which part of the plant is typically green and helps make food?

a) Leaf

b) Root

c) Stem

d) Flower

ANSWERS

1. d) All of the above

2. c) Root

3. b) Photosynthesis

4. c) Carbon dioxide

5. c) The Sun

6. d) All of the above

7. a) Lungs

8. b) Tree

9. b) Zoologist

10. c) Mars

11. b) Diamond

12. b) Pollen

13. c) Root

14. c) Amphibian

15. b) Heart

16. d) All of the above

17. b) Six

18. c) Mercury

19. c) 100°C

20. b) Caterpillar

21. b) Nitrogen

22. b) Condensation

23. b) Eyes

24. b) Revolution

25. c) Saturn

26. d) Combustion

27. b) Herbivores

28. c) Iron

29. b) Electrical energy

30. c) Fish

31. b) Kitten

32. c) Skeleton

33. b) Oxygen

34. b) Venus

35. c) Skin

36. b) Spine

37. d) Reptile

38. d) Nucleus

39. d) Flower

40. c) Orbit

41. d) Jupiter

42. c) Make food

43. b) Botanist

44. a) 0°C

45. c) Cell

46. c) Fossil fuels

47. c) Atmosphere

48. b) Bird

49. a) Lava

50. d) Smell

51. c) Sun

52. c) Chemical energy

53. b) Geologist

54. d) Jupiter

55. b) Earth

56. b) Spine

57. b) Lion

58. b) Transpiration

59. c) Gravity

60. c) Support the plant

61. b) Venus

62. b) Brain

63. b) Milky Way

64. c) Carbon dioxide

65. a) Tadpole

66. b) Chameleon

67. c) Absorb water and minerals

68. b) Nitrogen

69. c) Omnivores

70. c) Igneous

71. c) Stapes

72. c) Jupiter

73. c) Mammal

74. b) Hearing

75. a) Plasma

76. b) Pump blood

77. b) Spider

78. b) Elephant

79. a) Water cycle

80. c) Puppy

81. c) Rock

82. c) Reproduce

83. c) Solar energy

84. b) Jupiter

85. c) Meteorologist

86. b) Filter waste

87. c) Snake

88. c) Astronomy

89. d) Flower

90. c) Mammal

91. b) Carry oxygen

92. a) Atom

93. c) Stomach

94. c) Moon

95. b) Snow

96. b) Carnivores

97. b) Tongue

98. c) Sedimentary

99. b) Fight infection

100. a) Leaf

Printed at Repro India Ltd.